Contents

Safety

1. Acids must always be used in a well-ventilated area.
2. Avoid breathing any fumes from acids.
3. Use a fume cupboard when directed by the teacher.
4. Use acids carefully and wash off any splashes or drips immediately, using plenty of water.
5. Always add acid to water, not vice versa.
6. Always wear goggles when using the buffing wheel.
7. Keep both hands behind the cutting edge when using sharp tools.

Introduction

The word jewellery means personal body adorn- [illegible] includes rings, bracelets, necklaces, chokers, ear-rings, hair-slides and tiaras, as well as wrist straps, belts and buttons. The impact jewellery makes depends, to a great extent, on the design and quality of the craftmanship that has gone into it.

The first part of this book covers the making of jewellery from wire and sheet metal, and also explains the techniques of enamelling and etching. The second part covers techniques for casting metal. When you have decided what you would like to make and have chosen a method and materials, think carefully about your ideas. Always discuss them with your teacher before you start work.

Keep all your sketches, notes and sources of design in your project folder or workbook. Read and observe the safety notes and ask your teacher to clarify any points you do not understand. **Remember accidents can always be prevented.**

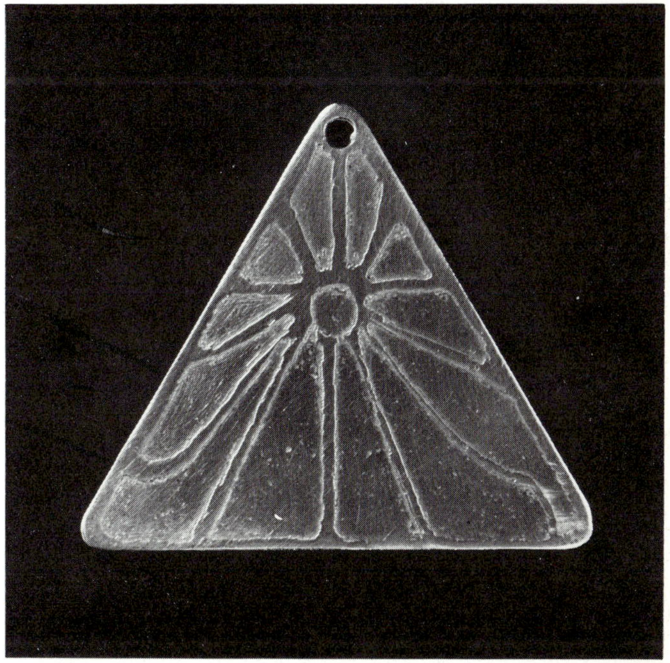

Design considerations

When designing jewellery it must be remembered that the finished article is to be worn and should be strong enough to withstand everyday use. It should be comfortable to wear, so sharp points and rough edges should be avoided. You will find when designing ear-rings, bracelets, necklaces, rings, belts etc., that different factors will have to be taken into consideration.

These problems are best solved by making detailed drawing or models and carrying out experiments which you can discuss with your teacher.

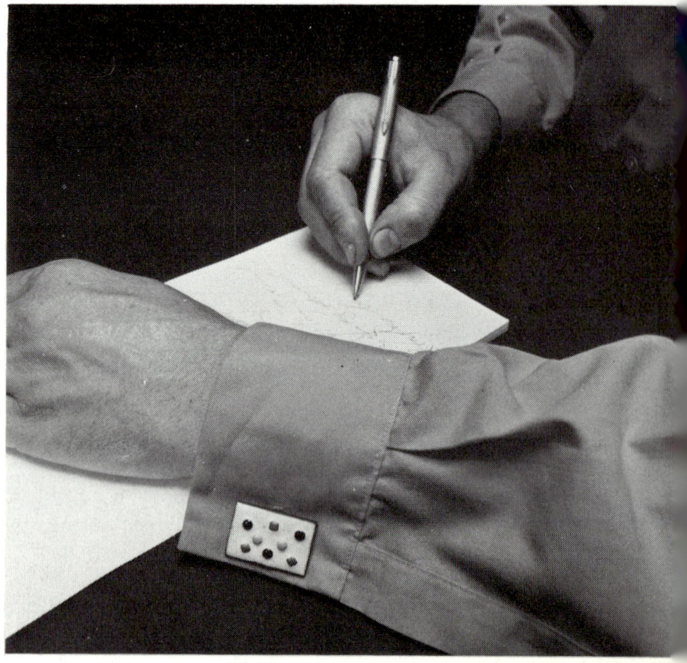

5

Wire jewellery

Wire can be shaped by beating, bending and twisting, to form very attractive jewellery. You will find that wires of gauge 2, 1.5 and 1 mm are the most suitable for making this type of jewellery.

Two examples are illustrated in colour on Design Source Card 7.

The photograph of a meandering river on page 5 was used to inspire these pieces of jewellery.

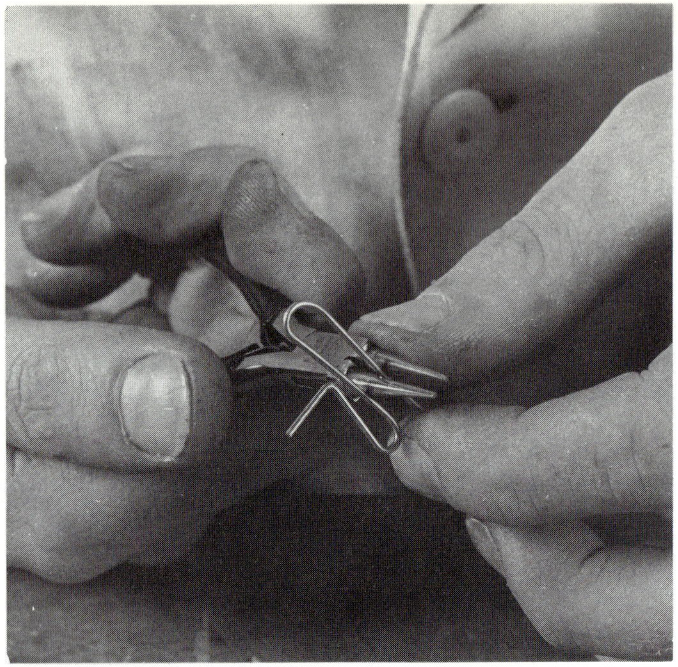

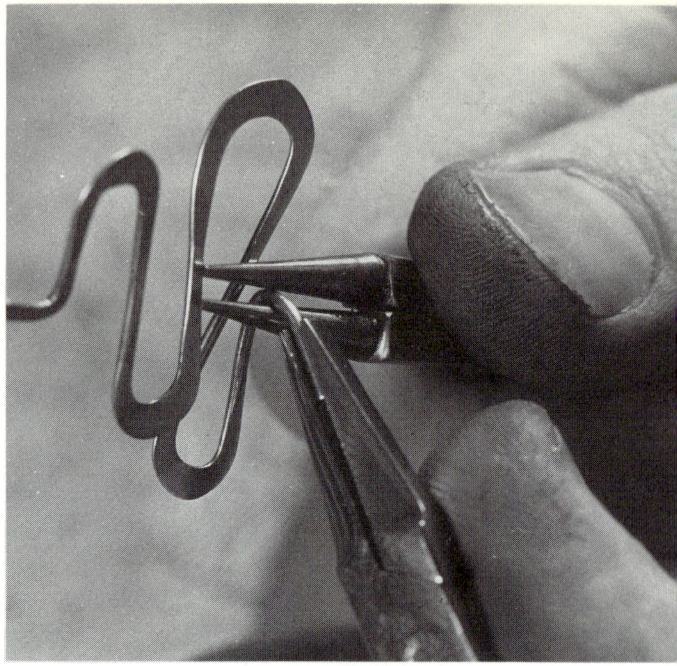

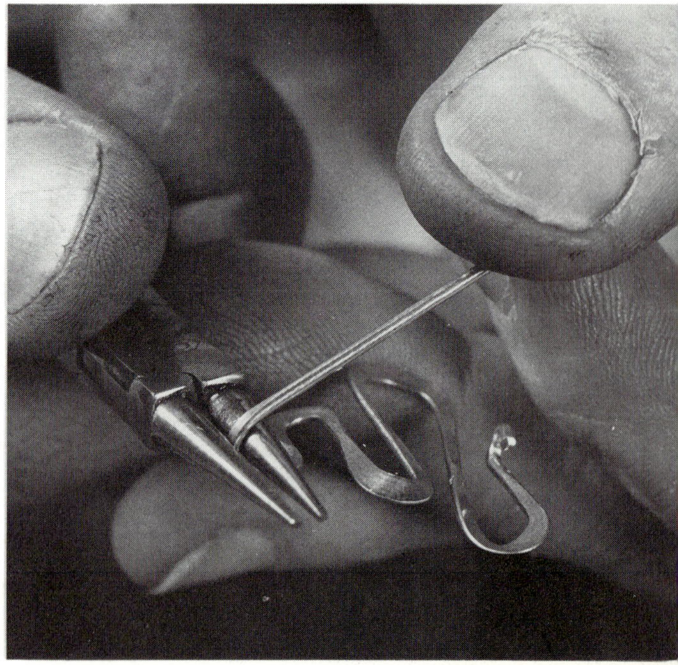

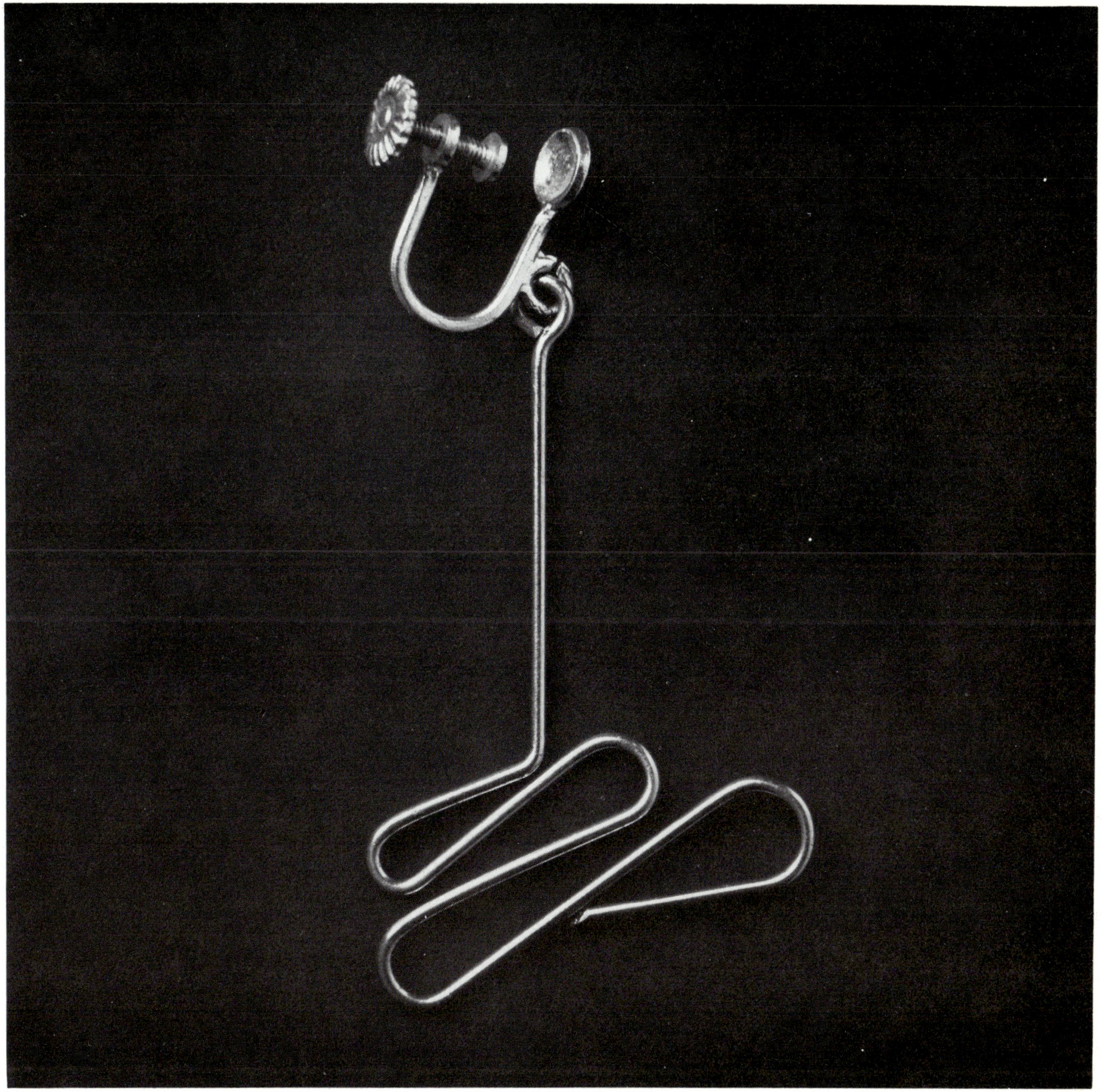

Simple wire ear-rings

1 Each pin has been shaped with round-nosed pliers and the joint silver soldered. The end is flattened on a bick iron.
2 Wind the wire around a piece of dowel rod, held in a vice, and cut through the wire with a saw to form the rings.
3 These rings should be silver soldered to form complete circles.
4 Each ring is shaped and a hole is drilled for the pin.
5 When the final shaping of the ring is completed the pin is silver soldered into place. The ear-ring is finished and buffed. The top loop is formed and the ear-ring fitted to the finding.

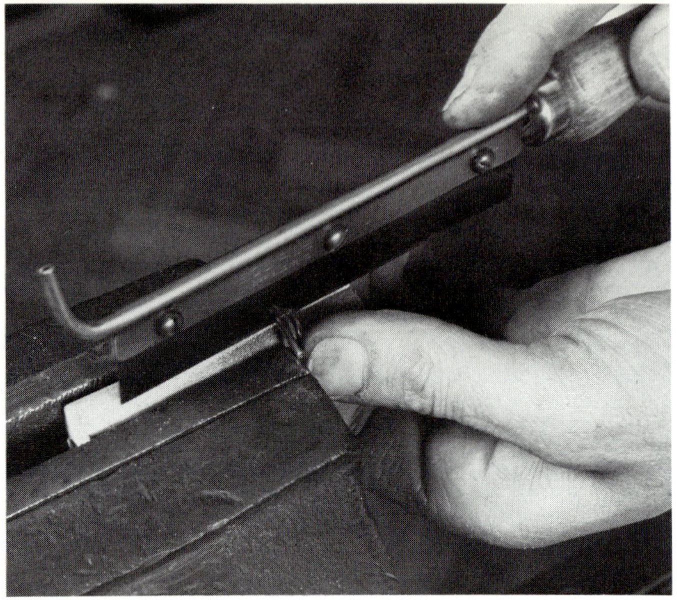

2. Cutting a ring.

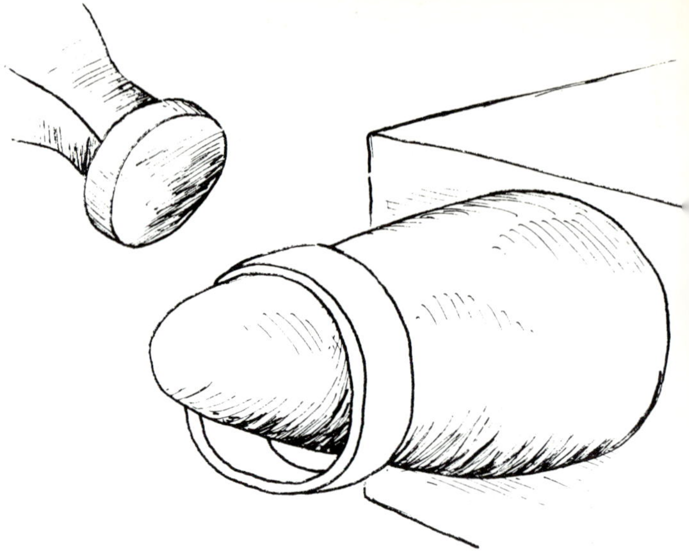

4. Each ring is shaped with a hammer and a bick iron.

1. Flattening the end of a pin.

3. The rings are soldered.

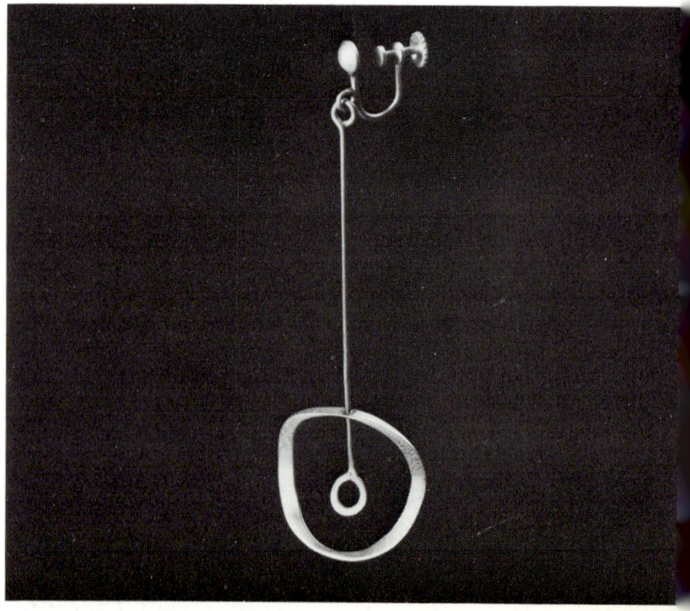

5. A finished ear-ring.

Sheet-metal jewellery

The pieces were cut out of a sheet of nickel silver using a piercing saw. They were joined together with hinges cut from small tube and silver soldered to each piece, as shown on the right.

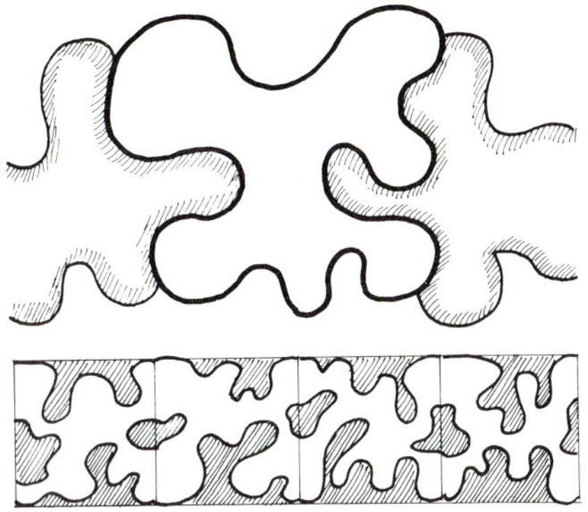

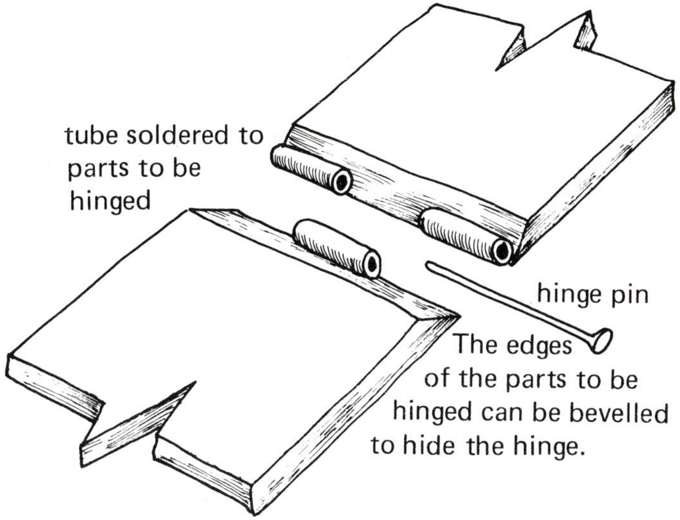

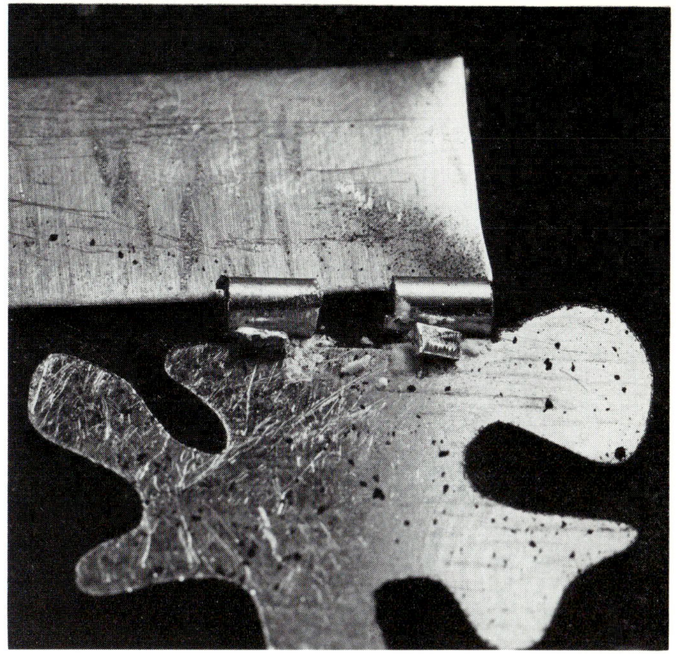

Design source: flower

Ring, pendant and choker made from nickel silver. They were cut out with a piercing saw and formed to shape. Simple stone settings were then silver soldered on and the stones mounted.

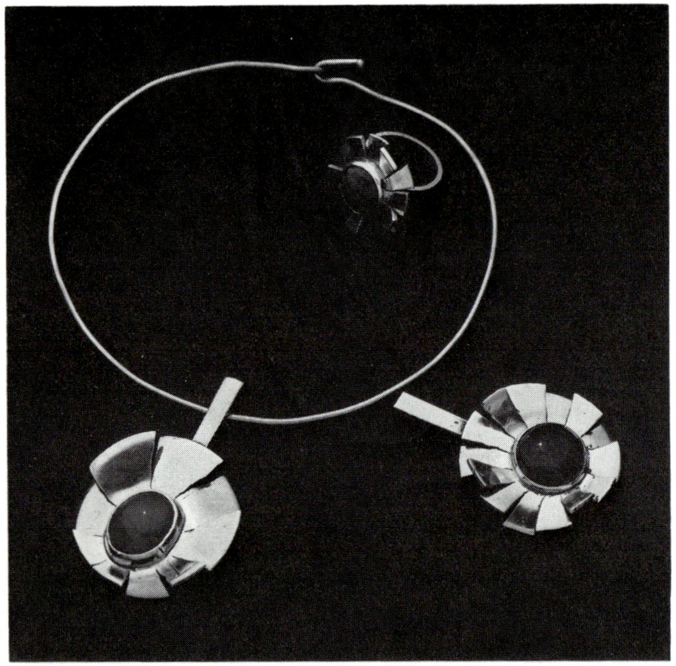

Design source: geometric

Bracelet made from nickel silver. Shapes were cut out with a piercing saw and drilled to take jump rings to join them together. Jump rings are small split rings which are not soldered together.

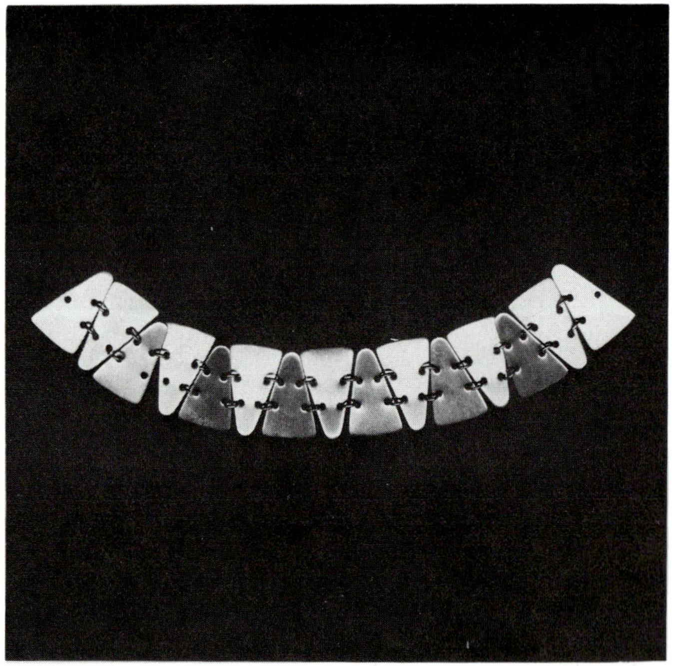

Design source: name

Choker made from nickel silver. The design was pierced out and joined with jump rings and a wire band.

Enamelling

Enamelling is the technique by which powdered glass is melted and fused onto the surface of metal. Examples of enamelling are baths, sinks, saucepans, kettles and cookers. The technique can also be used to make very colourful pieces of jewellery. Some examples are illustrated in colour on Design Source Cards 6 and 8.

Safety
Enamel can contain poisonous substances. Do not put your fingers in your mouth or rub your eyes while using enamel.

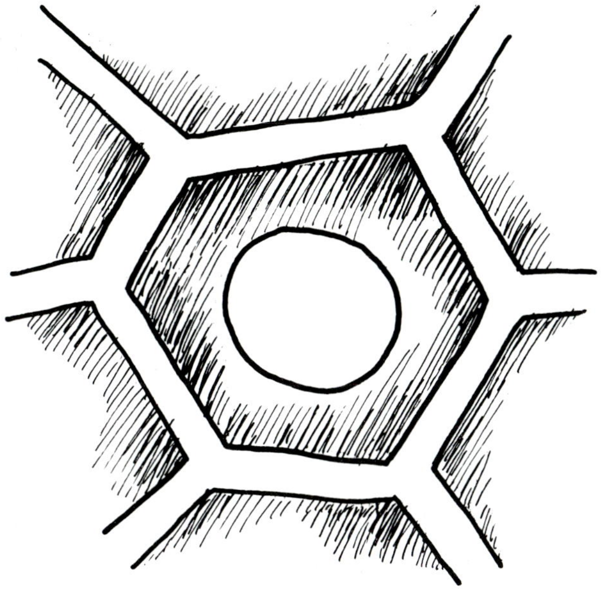

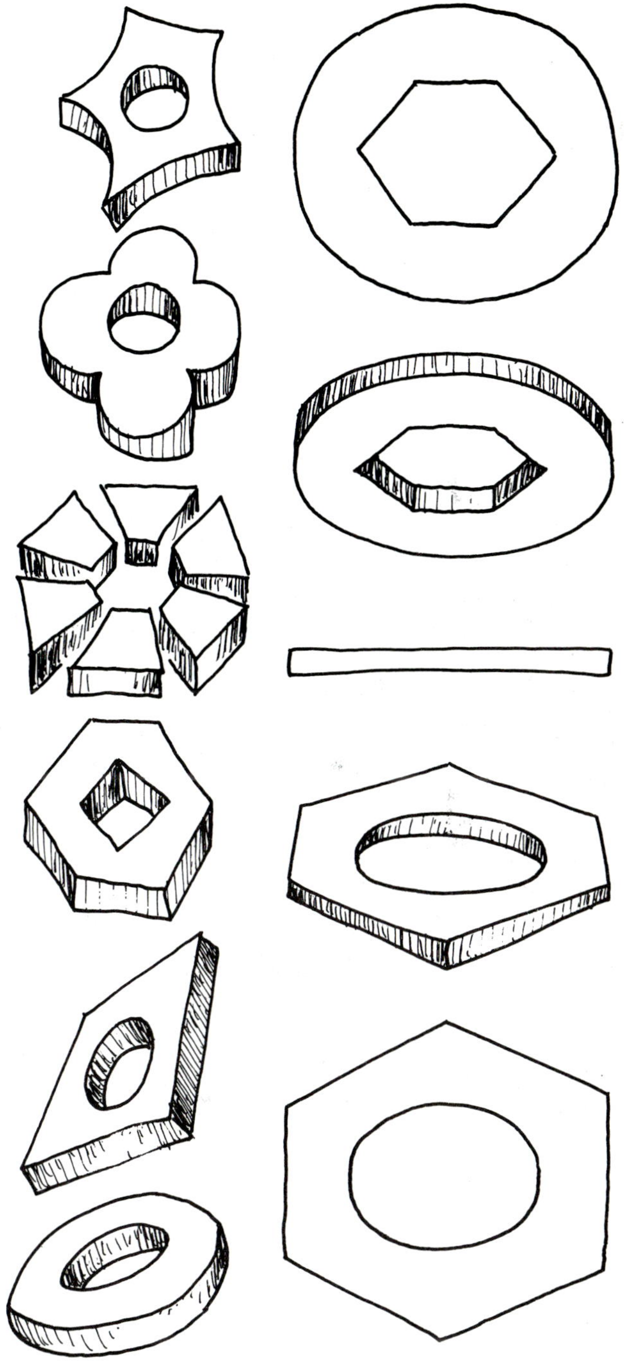

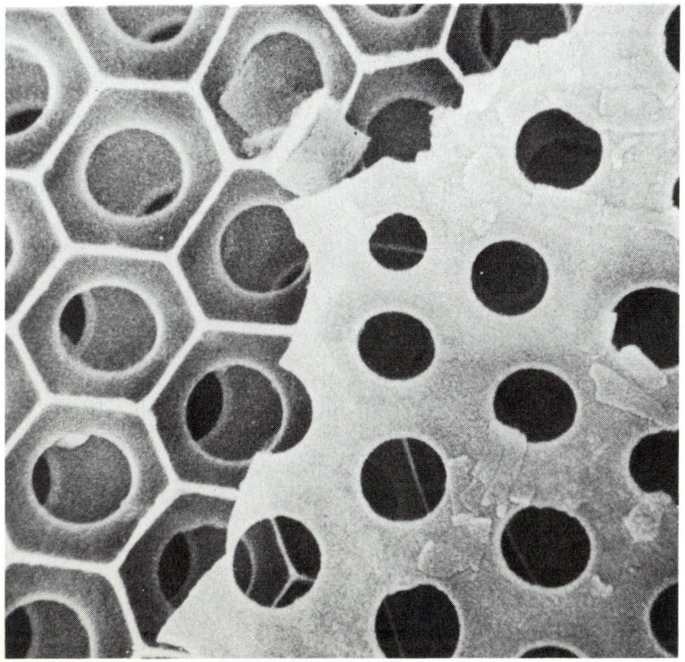

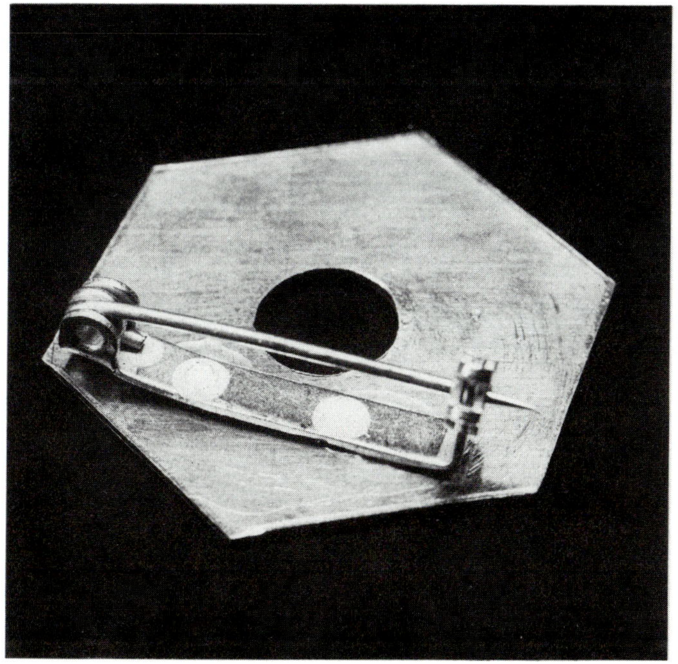

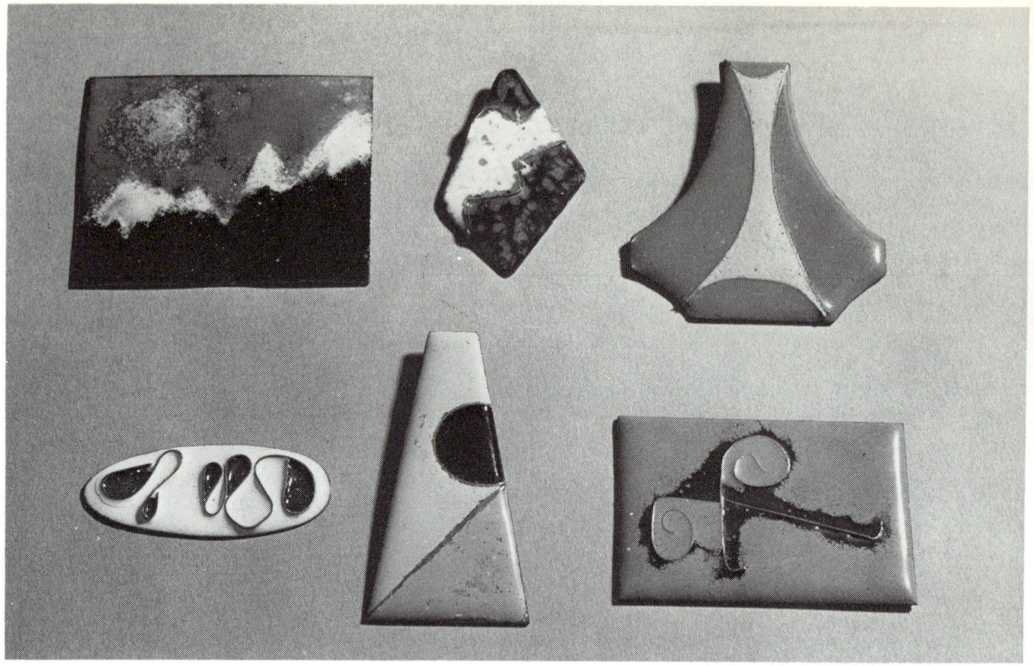

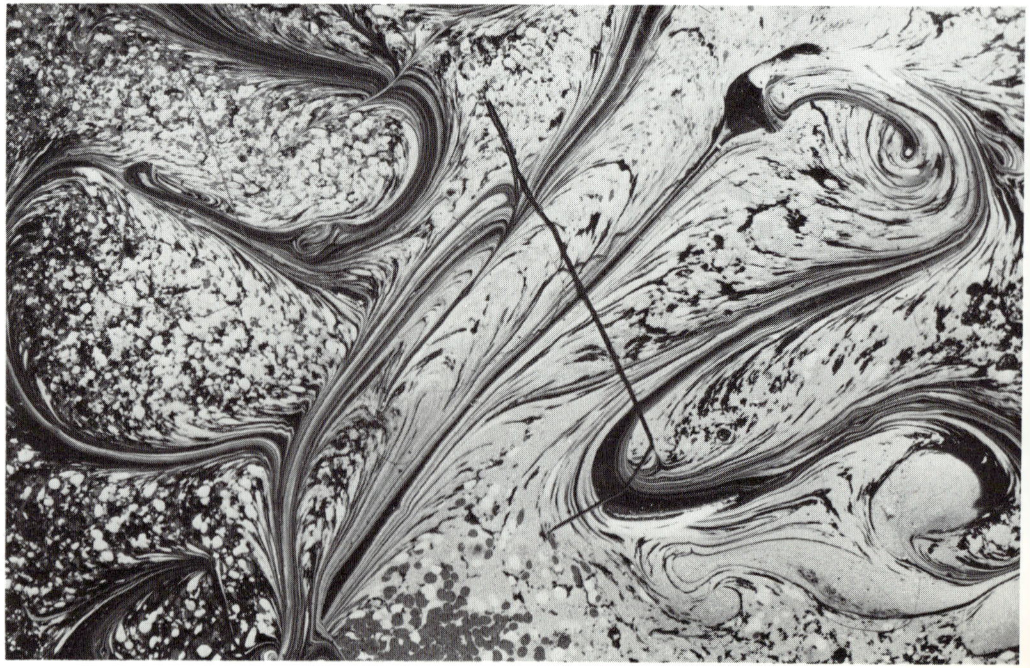

The process of enamelling

To enamel a piece of copper follow these instructions:

1 Transfer the design onto a piece of copper and cut out the shape.

2 File the edges smooth.

3 Clean the metal with steel wool to remove dirt and grease. When it is clean and shiny, rinse it in water.
Note: Carry the work with metal tongs to avoid leaving fingermarks on it.

4 Dry the metal and place it on a clean piece of paper before painting the surface with 'Polycell' paste.

5 Shake the enamel through a sieve onto the pasted surface.
6 Without disturbing the enamel place the work on a wire mesh trivet and place the trivet on top of the kiln to allow the paste to dry.
7 When the paste is dry place the work, still on the trivet, in the kiln. In about 2 minutes fusion will take place and the enamel will become smooth and shiny. Remember to wear oven gloves when putting work in and taking it out of the kiln.
8 Remove the trivet and work from the kiln and allow to cool slowly. Slow cooling will avoid any cracking of the enamelled surface.
9 When the work is cool, remove it from the trivet and place it in an acid bath using brass or copper tongs. The acid will remove any black oxide which may have formed on the metal during firing.
10 After a few minutes remove the work and wash it in water.

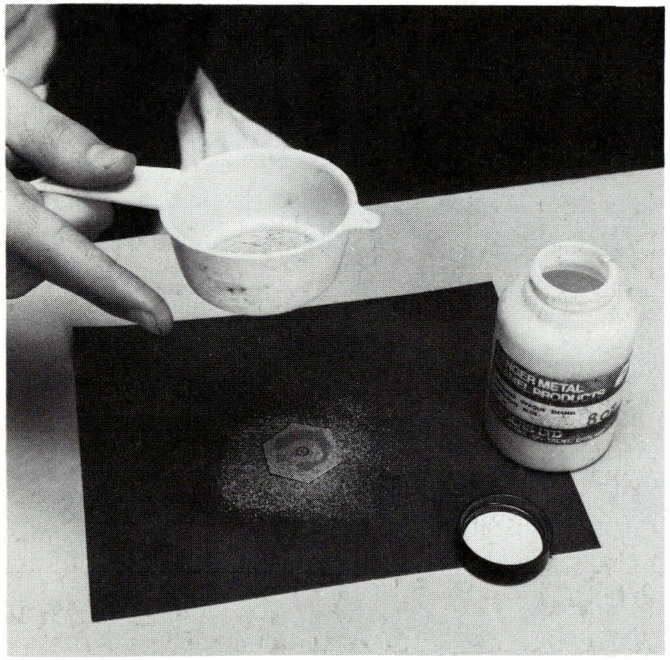

Etching

When a piece of metal is etched some of it is burned away by acid. A textured effect is achieved by the contrast between the polished surface of the metal and the dull, rough surface which has been burned by the acid.

1 Make sketches of your ideas.
2 When you are satisfied with your design make an exact drawing on a piece of paper.

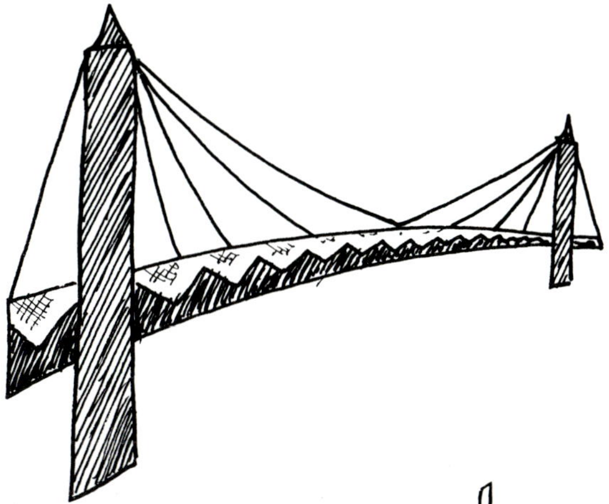

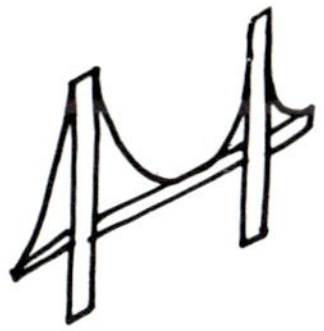

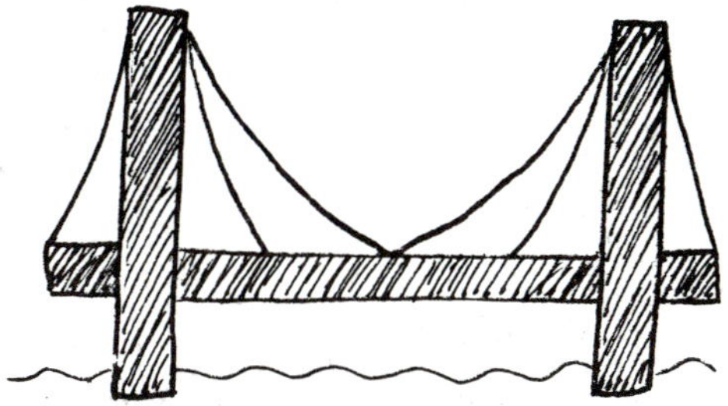

3 Transfer your design onto the copper and etch it, as described on page 16.
4 Fit findings as required.

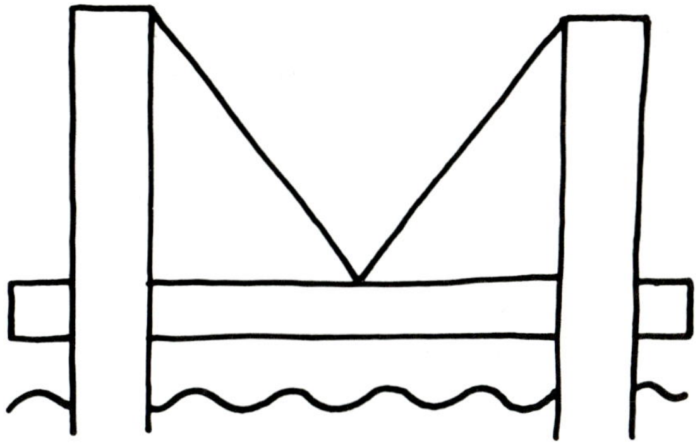

The process of etching

Safety

Always follow the safety precautions on page 3 when using solutions for etching. Your teacher will advise you.

To etch a piece of metal such as copper, brass, gilding metal etc., follow these instructions:

1 Clean the surface of the metal with pumice powder or wire wool.

2 Choose a design. Place a piece of carbon paper over the metal. Place the design on top of the carbon paper, and trace it directly onto the metal.

3 Areas not to be attacked by the acid must be painted with stopping-out varnish; the edges and the back must also be covered. Allow to dry.

4 Place the work carefully in the acid bath. Make sure that all the work is covered by the acid solution. When bubbles appear on the surface of the metal, remove them with a feather, or the work will be unevenly etched. **If you are using nitric acid for etching you should carry out this stage in a fume cupboard.**

5 Consult with your teacher about when to remove your work from the acid bath. Be sure to use a pair of brass tongs. Then wash the work in running water.

6 Remove the stopping-out varnish from the work with a suitable solvent.

7 Finish off the work by polishing it on a buffing wheel.

Casting

This section covers the making of metal jewellery by casting techniques. These are more advanced methods of making jewellery and you should not use them for your first piece of work.

Making jewellery using the 'lost-wax' casting process involves the following basic procedure:

1. The model is made in wax.
2. The model is set in heat-resisting plaster.
3. The wax model is burned out, 'lost' and then the molten metal is forced into the mould cavity in a centrifugal casting machine or a vacuum casting machine.

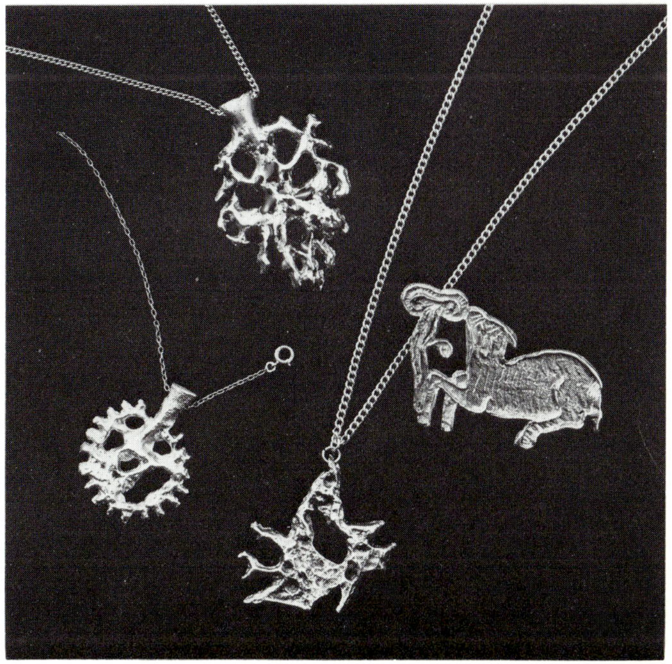

Design process

Design source

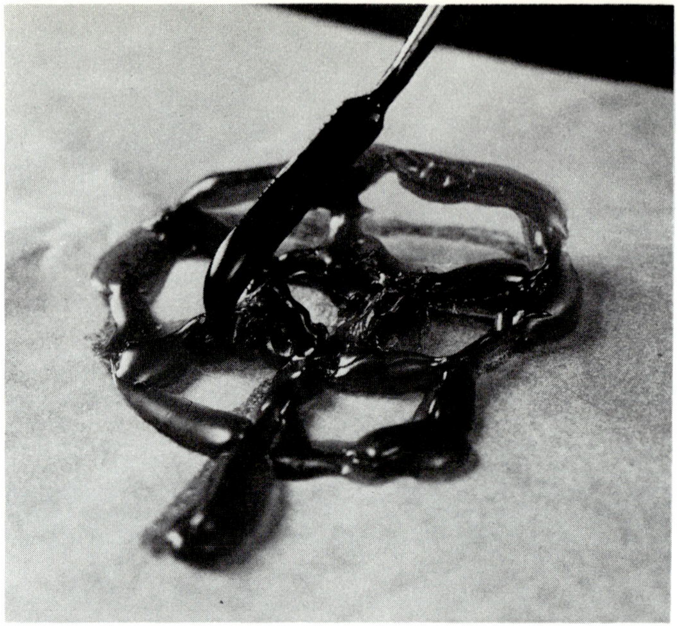

Modelling the wax

Vacuum casting

Developing ideas

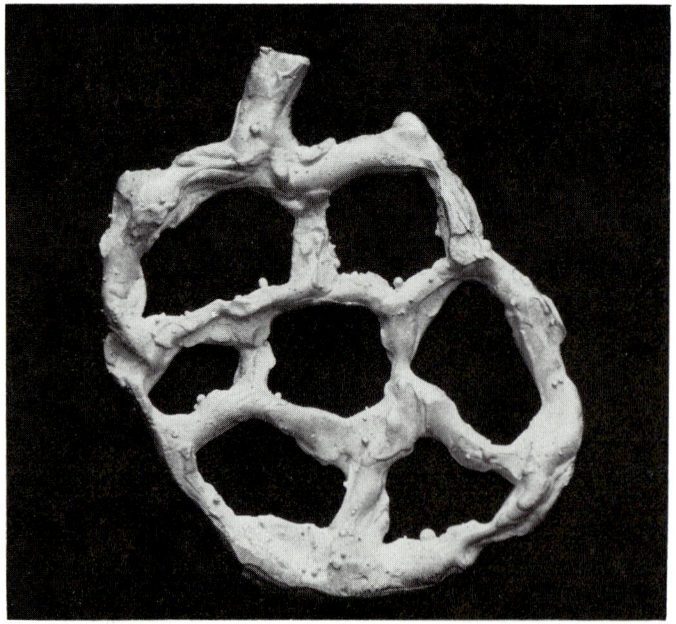

Finished model

The finished work

Wax

Wax modelling is unique in that it is the only process for making jewellery that enables us to mould a shape with our fingers and then have this shape perfectly cast in metal. It has the great advantage that we can feel and see what the finished piece of jewellery will be like before it is cast.

Two different types of wax can be used, and both can easily be melted. **Hard wax** is bought specially for this purpose and can be cut, filed, drilled or scraped. The other type of wax available is **soft wax** and is usually available as candle or beeswax. It can be moulded by hand, bent to shape by immersion in warm water or passed through a gentle flame. Soft wax can also be pierced with a hot piece of metal.

Making wax melts
This photograph shows melted wax being poured down the side of an aluminium sheet, and shows some of the interesting shapes which result. These shapes can be joined together to make models for casting. The advantage of using an aluminium sheet is that cold wax does not stick to it and the wax 'melts' can easily be taken off.

This is an example of a pendant cast from a pattern made from candlewax. The ribs were made by cutting slits in the wax, twisting them slightly and pinching to thin the edges. The idea for the design came from a dried and torn lettuce leaf.

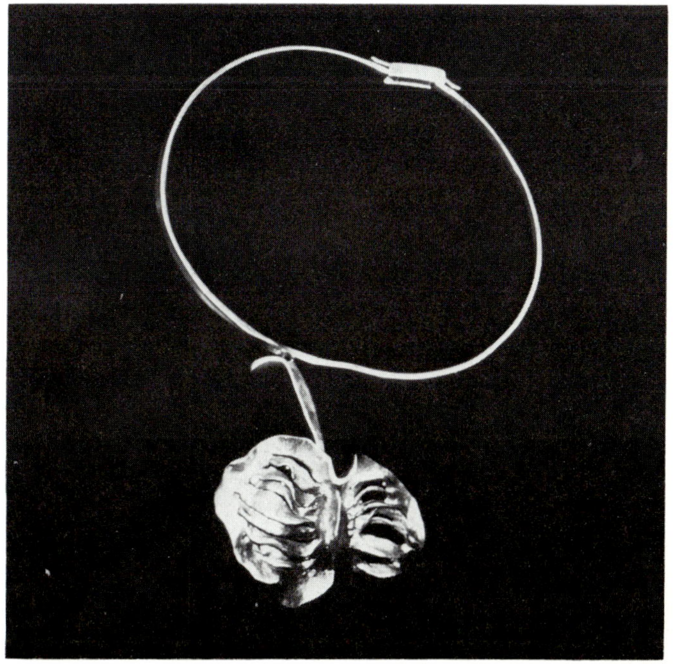

Wax modelling
Wax models or patterns can be made with soft beeswax or candlewax. These photographs show how easy it is to soften and mould the wax to a variety of simple but interesting shapes.

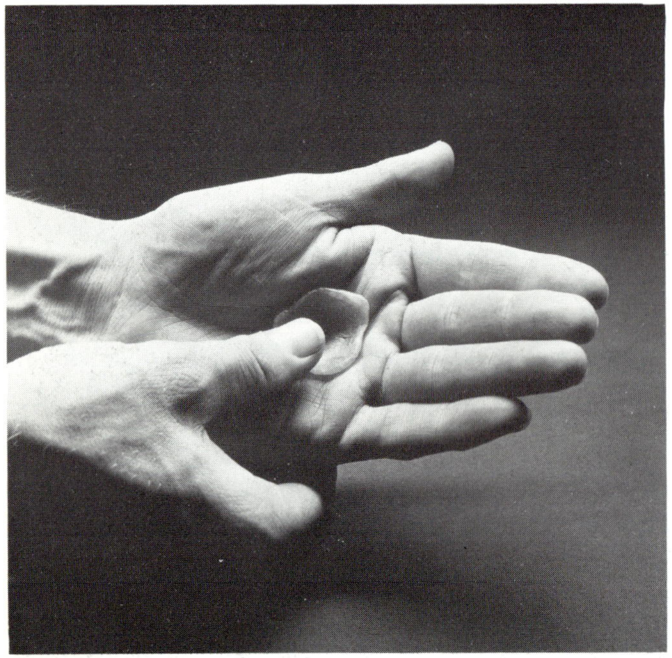

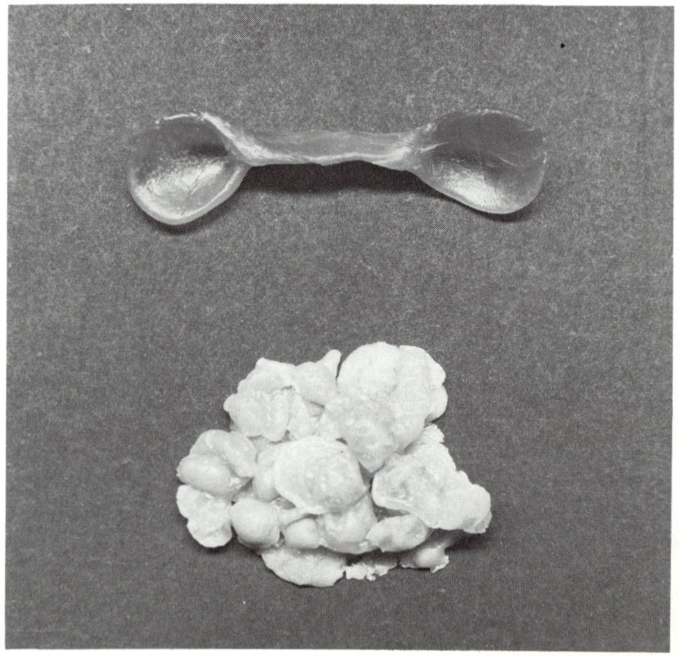

Wax sheets

Wax sheets can be bought but it is simple and much cheaper to make your own to whatever thickness you desire. You first melt some wax in a ladle and then pour the hot wax into a tin lid or aluminium tray.

Wax forms cut from wax sheets.

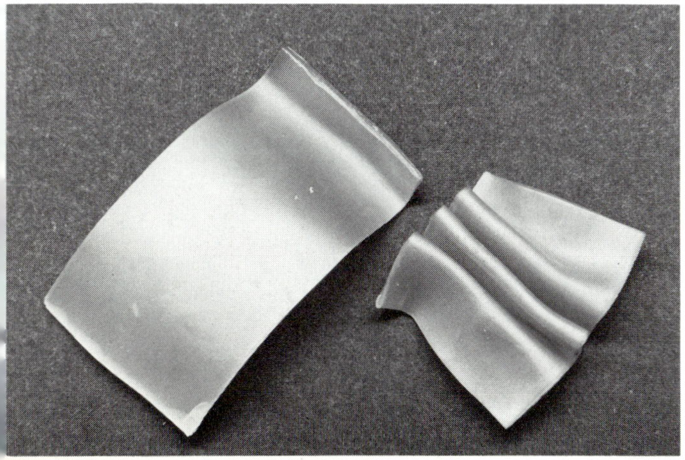

Shapes made by manipulating wax sheets.

Wax forms

Wax forms can be bought already shaped but you can also cut and shape your own to make very interesting pieces of jewellery. First cut the form you want in a sheet of wax and then shape it after immersing it in warm water. Separate shapes can then be joined together by using more melted wax. Interesting shapes can also be made by manipulating sheet wax.

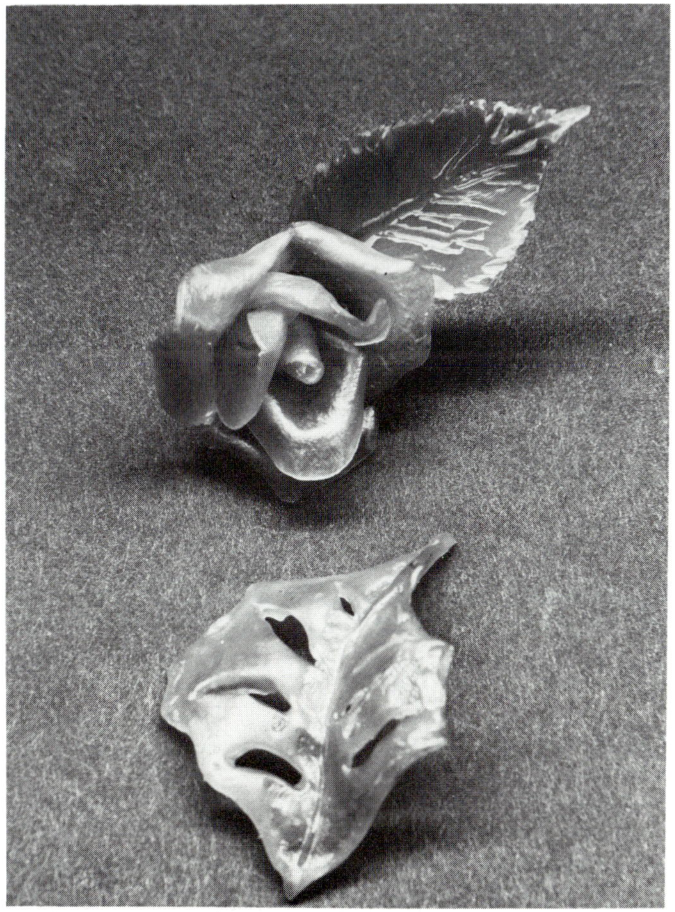

Examples of wax forms that can be bought.

Wax wires

Wax wires can be bought; these are some of the shapes available.

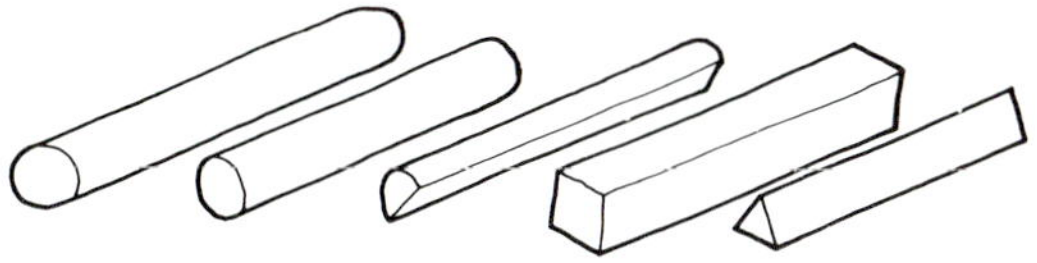

Wax wires can be turned around a piece of wooden dowel rod and cut into rings. This is a similar method to the one used for cutting rings from metal wire (page 8).

Interesting shapes can be made by joining wax wires and melting balls of wax on the end of them. Wax strips can be bought and then cut with scissors to make wires of your own width and thickness.

Here is an example of a brooch made from wax wires of different sections and sizes in which extra blobs of wax have been added to give the joints more strength.

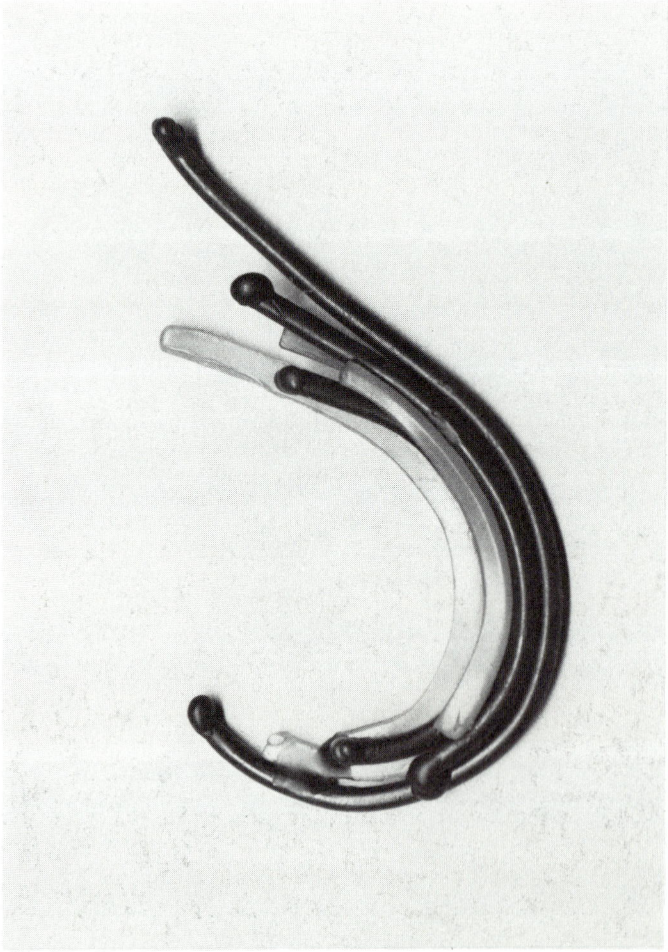

Wax blocks

Wax blocks can be bought and then carved by drilling, piercing or filing them to shape.

1 Draw a design on tracing paper.

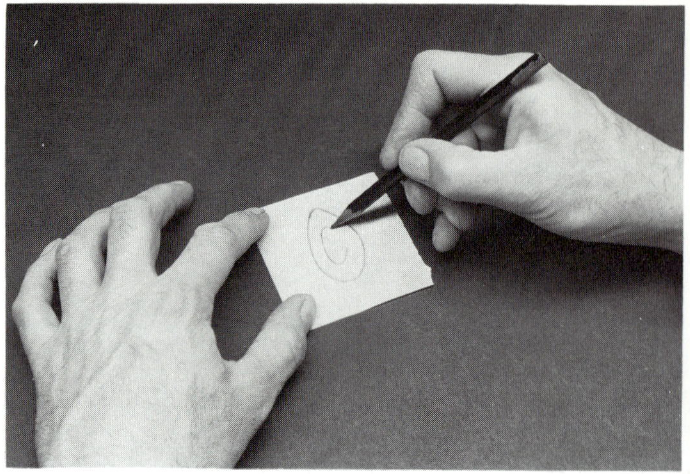

2 Stick the design on to the carving wax and prick through it into the wax.

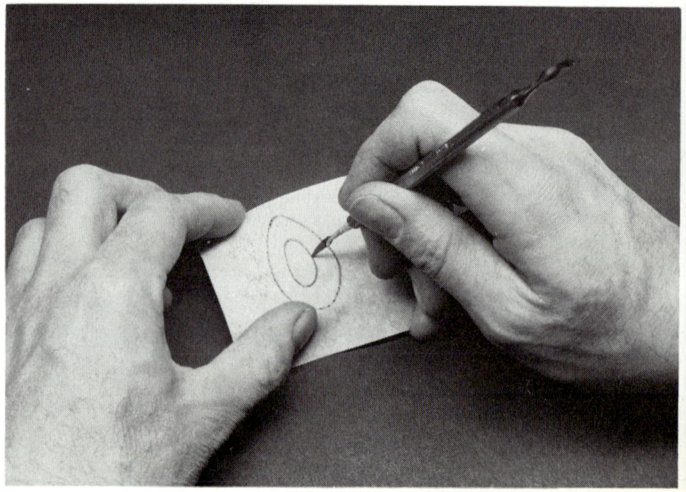

3 Rub talcum powder into the pin pricks and your design will be easy to see.
4 Saw around the outline, and then drill a hole through the centre and pierce it out.
5 File inside the outline so as to round out the edges at the angle you want.

The modelling spatula

The photographs show how to make a wax model for a ring using melted wax.

1 Wrap a piece of tracing paper round a metal rod and secure with a blob of wax or Sellotape.

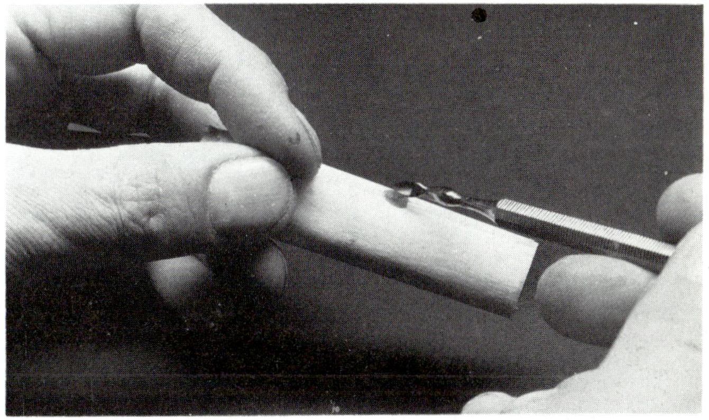

2 Heat the tip of the modelling spatula in a flame, then touch the wax with the hot tip of the spatula and pick up a small blob of wax.

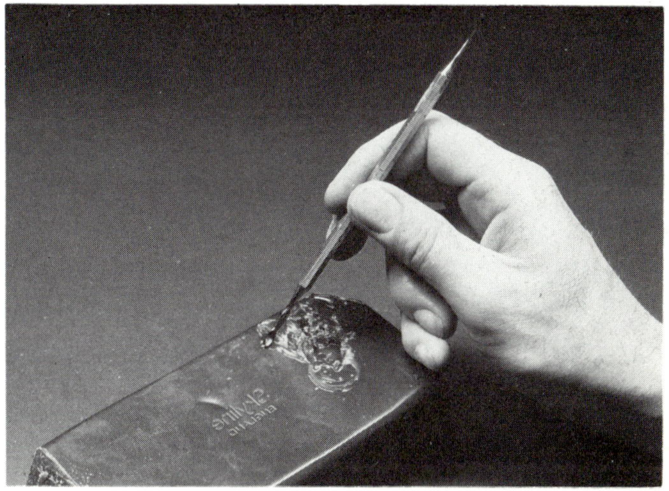

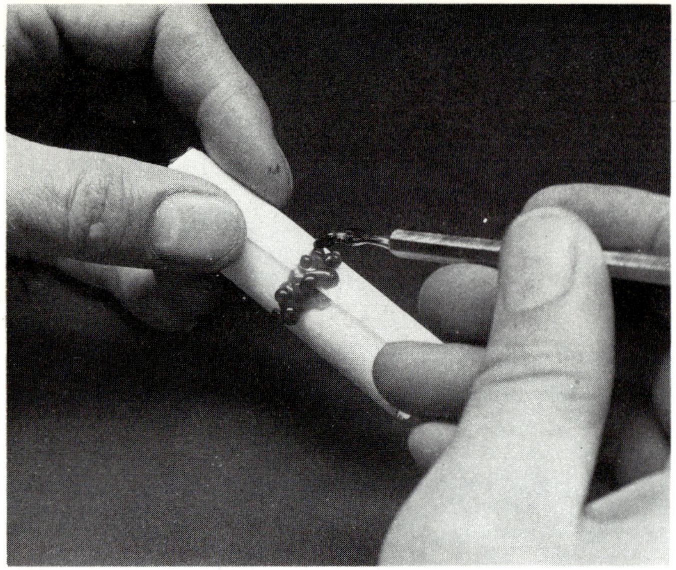

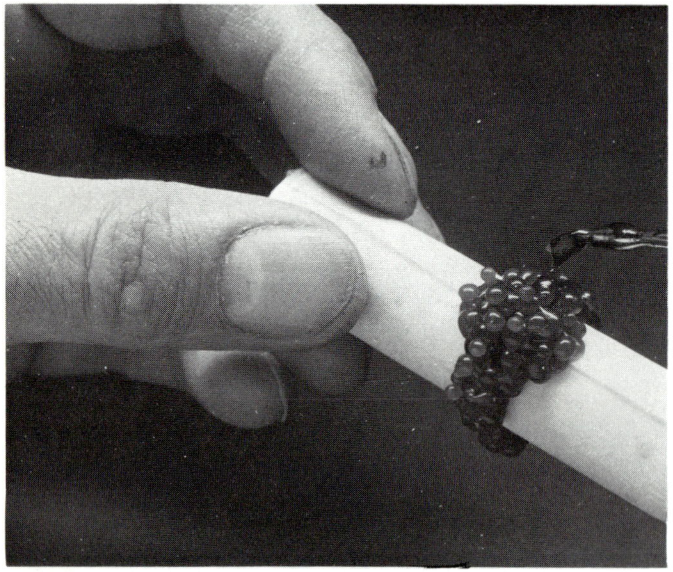

3 Build up the ring with a series of blobs. Make sure all the blobs are melted into each other by gently 'pulling' a little wax bridge between each blob. You will need to heat the tip of the spatula before doing this.

4 When the model is finished carefully tear off the paper to remove the model from the rod. If the model is still stuck it can be released from the metal rod by gently heating the handle end and twisting the wax model very gently until it is released from the rod.
Note: Use as little heat as possible or you may damage your model.

If this method of releasing a model proves difficult take a piece of aluminium foil, wet it a little on one side, and wrap it tightly round the rod to build your model on. The foil will easily slide off the rod when you have finished and is then pulled gently away from from the inside of the model.

Trailing wax

1 Draw a basic design.

2 Cut the design out of sheet wax using the spatula.

3 'Trail' the wax as shown to build up your model. Each time you add a small blob of wax (not too hot) pull out one side of it, and blow gently as you pull ('trail') so as to cool and set the wax.

Design ideas

Design ideas can be obtained from nature and further unusual shapes achieved by experimenting with the modelling wax.

You should get into the habit of jotting down your idea on paper whenever you come across a new one.

1 Holes can be melted in wax with a hot spatula.
2 Cut up various sections of wax and add them to blobs. A model for jewellery could be made from this arrangement.
3 Bend a ring shape in wax and add wax wires to form a holding device for a stone.
4 Casting wax in water:
Melt some wax in a spoon or ladle then slowly pour the molten wax into cold water 10–20 mm deep.
Remove the shape from the water, insert a needle into the centre and pour off any excess wax under the hardened surface.

Note: Different depths of water can give different shapes. There are many other experiments and you should try some of them for yourself.

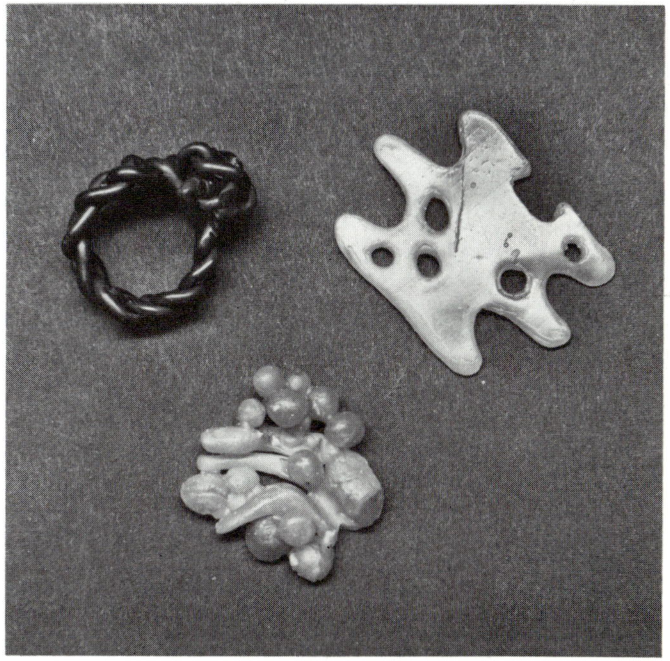

The casting process

When your wax model is complete it has to be sprued before it can be embedded in investment, which is casting plaster. Sprues are wax rods which are attached to the model. They support the wax model while the investment is being poured onto it as well as providing escape channels for the wax which has to be removed before the molten metal is forced, through the same channels, into the cavity. The photographs show the model ready for investing, and a vacuum casting machine.